LITTLE
QUICK FIX:

CHOOSE YOUR
METHODOLOGY

#LittleQuickFix

Sara Miller McCune founded SAGE Publishing in 1965 to support the dissemination of usable knowledge and educate a global community. SAGE publishes more than 1000 journals and over 800 new books each year, spanning a wide range of subject areas. Our growing selection of library products includes archives, data, case studies and video. SAGE remains majority owned by our founder and after her lifetime will become owned by a charitable trust that secures the company's continued independence.

Los Angeles | London | New Delhi | Singapore | Washington DC | Melbourne

LITTLE QUICK FIX:

CHOOSE YOUR METHODOLOGY

Charlotte
Jane Whiffin

Los Angeles | London | New Delhi
Singapore | Washington DC | Melbourne

Los Angeles | London | New Delhi
Singapore | Washington DC | Melbourne

SAGE Publications Ltd
1 Oliver's Yard
55 City Road
London EC1Y 1SP

SAGE Publications Inc.
2455 Teller Road
Thousand Oaks, California 91320

SAGE Publications India Pvt Ltd
B 1/I 1 Mohan Cooperative Industrial Area
Mathura Road
New Delhi 110 044

SAGE Publications Asia-Pacific Pte Ltd
3 Church Street
#10-04 Samsung Hub
Singapore 049483

Editor: Alysha Owen
Editorial assistant: Lauren Jacobs
Production editor: Rachel Burrows
Marketing manager: Ben Sherwood
Cover design: Shaun Mercier
Typeset by: C&M Digitals (P) Ltd, Chennai, India
Printed in the UK

Library of Congress Control Number: 2020939958

British Library Cataloguing in Publication data

A catalogue record for this book is available from
the British Library

ISBN 978-1-5297-2971-9

At SAGE we take sustainability seriously. Most of our products are printed in the UK using responsibly
sourced papers and boards. When we print overseas we ensure sustainable papers are used as measured
by the PREPS grading system. We undertake an annual audit to monitor our sustainability.

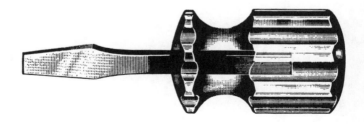

Contents

Everything in the book! 4

Section 1   What is methodology? 9

Section 2   Why does methodology matter? 25

Section 3   How do I choose between
             quantitative, qualitative and
             mixed methodologies? 37

Section 4   When should I use quantitative
             methodology? 53

Section 5   When should I use qualitative
             methodology? 65

Section 6   When can I mix qualitative and
             quantitative methodologies? 79

Section 7   DIY: How do I choose the right
             methodology for my research? 93

             Glossary 106

2 MIN summary

Everything in the book!

Section 1 Methodology is often used differently by different authors and is therefore commonly misunderstood. Here we use it to describe the thinking behind how the study is put together.

Section 2 Choosing a methodology is very important as this decision will guide all other steps in the research process. If decisions are made correctly the study will achieve methodological congruence.

Section 3 Researchers choose from qualitative, quantitative and mixed methodologies. No methodology is better than another; there is simply the right approach for the research question.

Section 4 Quantitative studies are usually big, replicable and objective, and measure the phenomena of interest using statistical analysis to generate results.

Section 5 Qualitative studies are usually small, subjective, exploratory designs which aim to interpret, understand and illuminate the subjective experiences of participants.

Section 6 Mixing methodologies enables researchers to combine qualitative and quantitative approaches within one study, when the research question cannot be answered in full by using a quantitative or qualitative approach only.

Section 7 DIY: To choose your own methodology you need to understand the research problem, the research question and yourself as a researcher.

Section 1

What is methodology?

summary

The fundamental assumptions that guide decision making in a study. As such, methodology is the glue that holds everything together.

All researchers need methodology

When designing a study all researchers make decisions about methodology. These decisions are informed by many different and complex theoretical ideas about what is important in the world around us, and how problems can and should be investigated through scientific inquiry. However, once a methodology has been selected as appropriate to address the aims and objectives of a study its underpinning theory can then drive all other decisions in the research process. Methodology, then, is the fundamental assumptions that guide decision making and, as a core influence on the study, **methodology is the glue that holds everything together**. This glue, if used correctly, will support the researchers to make appropriate decisions throughout the research process.

WHY AM I CONFUSED ABOUT METHODOLOGY?

First of all, don't be so hard on yourself. Even experienced researchers find methodology confusing. With a subject like methodology it sometimes feels like the more you read the less you understand. This is very normal. In my experience, the most common reason students are confused about methodology is because authors do not use research terms consistently, either in academic textbooks or published papers. In addition, many published papers do not discuss methodology at all as part of an explicit stage in the research process. Therefore, not only can the definitions of methodology be confused and contradictory but their application is often hidden away from view.

DEFINING METHODOLOGY

It is worth mentioning here that there are other terms which are similarly poorly defined in research. Inconsistent definitions mean that the boundaries between methodology and other parts of the research process can become blurred. This textbook defines methodology as:

'the fundamental assumptions that guide decision making in a study'

This definition may not be the same as in other texts you might read. However, if you understand methodology as it is presented here it will help you apply methodology appropriately and consistently in the context of your own work and also help you to see how others have defined it in theirs.

WHAT OTHER TERMS SHOULD I LOOK OUT FOR?

There are a number of other terms that are relevant to any discussion of methodology. An understanding of these, and how they are defined in this text, will further help you understand methodology. These terms are:

- Paradigms
- Research design
- Methods

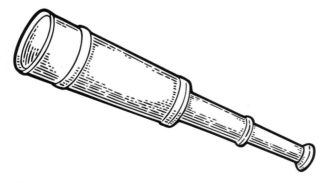

PARADIGMS

Paradigms are world views of what is influencing the conduct of the research. They are concerned with *epistemology*: the nature of knowledge and how knowledge can be gained, and *ontology*: the nature of truth and reality. Therefore, paradigms influence how the researchers see the research problem and what methods they should use to investigate it. Some examples of leading paradigms include those listed below but be aware there are lots of others.

- Positivism
- Interpretivism
- Pragmatism

Every paradigm has its own particular perspectives on knowledge and reality. These perspectives then inform what type of methodology is most appropriate to use. For example, the positivist paradigm defines knowledge as absolute and verifiable, and reality as objective and fixed. Selecting this paradigm would usually lead a researcher to use a quantitative methodology. The interpretivist paradigm defines knowledge as created through a process of interpretation and considers reality as subjective and multiple. Those who choose an interpretivist paradigm should use a qualitative methodology. The paradigm of pragmatism defines knowledge as pluralistic and reality as both singular and multiple. Pragmatism is the paradigm usually used by those who want a mixed methodology study.

RESEARCH DESIGNS

Research designs are particular styles of research. For example, if you want to bake a cake, a simple recipe of flour, eggs, sugar and butter will create a nice-tasting sponge but if you want to create a Black Forest gateau, Battenberg or Bakewell tart, there are special ingredients and specific instructions for how these should be combined if the end product is to achieve its goal. This is the same in research. The list below presents just some of the most common research designs you will come across and in each there are core ingredients and important steps which, if actioned appropriately, will lead the researcher to the intended result.

- Grounded theory
- Ethnography
- Phenomenology
- Narrative
- Case study
- Randomised control trial
- Cohort
- Case control
- Mixed

METHODS

Methods are the techniques required in a study to carry it out. They relate to who the researchers should recruit, how they should collect data and how the data should be analysed. Therefore, methods are the operational procedures required to achieve the aims of the study.

Some examples are listed below.

- Sampling
- Questionnaires
- Triangulation
- Interviews
- Analysis

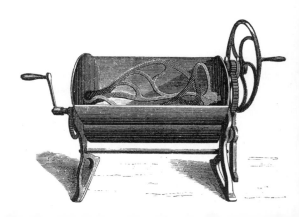

WHAT TYPES OF METHODOLOGY ARE THERE?

There are three main types of research methodology: qualitative, quantitative and mixed. Quantitative methodology is interested in question like How many? How often? How much? Qualitative methodology is used to investigate experience, context and culture. Mixed methodologies are useful when relying only on a qualitative or quantitative approach will not fully answer the research question. Each methodology will be examined in more depth in subsequent sections.

Quick reminder! At this point it is worth reminding you about the lack of consistency around the term 'methodology' – some authors define the terms 'qualitative' and 'quantitative' as paradigms. In addition, the research designs listed earlier are also defined by some authors as methodologies while others would call them methods. See why it's confusing? However, if each term is understood by the classification listed in the paragraph above it will help you understand how they all fit together within the research process.

WHERE DOES METHODOLOGY FIT IN THE RESEARCH PROCESS?

The research process is a linear stepwise progression of decision making through a study although in reality the process of designing a study is often more iterative than sequential. Ask anyone who has tried! At the top of the decision tree, concepts are more abstract and philosophical, at the bottom more applied and pragmatic.

- **What's your paradigm (epistemology and ontology)**
  Positivism, interpretivism, pragmatism

- **What's your methodology?**
  Quantitative, qualitative, mixed

- **What's your research design?**
  Grounded theory, ethnography, phenomenology, narrative, case study, randomised control trial, cohort, case control, mixed

- **What are your methods?**
  Sampling, interviews, questionnaires, analysis, triangulation

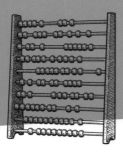

Typically, decisions are easier further down the research process because they are the practical things required in research to actually carry it out. In contrast, higher order constructs are more difficult to understand and apply but these are really important because they lead all subsequent decisions. For example, if you select positivism as the paradigm the methodology will be quantitative, and an observational design might be appropriate together with a large sample and statistical analysis.

Now let's see if you can correctly connect the description of epistemology and ontology with its associated methodology (quantitative, qualitative or mixed).

Fill in the blanks.

1 **Epistemology** (Knowledge is constructed, discovered) + **Ontology** (Reality is subjective and multiple) =

.... methodology

2 **Epistemology** (Knowledge is absolute, verifiable) + **Ontology** (Reality is objective and fixed) =

.... methodology

3 **Epistemology** (Knowledge is pluralistic) + **Ontology** (Reality is singular and multiple) =

.... methodology

2 Section

Why does methodology matter?

summary

It provides the instructions
for how to conduct a valid
piece of research and
interpret the data.

Methodology matters

Methodology matters for two reasons. This first is that **methodology provides the instructions for how to conduct a valid piece of research and interpret the data that is collected.** Without it, research would just be a combination of techniques with no alignment or synergy between them. In some ways doing whatever you like may seem quite liberating, but if research is to be a robust scientific endeavour it must proceed in a systematic way and rigorous methodology facilities this. The second, **given that research is a process of generating new knowledge, is that methodology helps us to understand** both *how* to create new knowledge, and *what* new knowledge is being created through the interpretation of the data.

A BRIDGE BETWEEN PARADIGMS AND METHODS

Methodology connects the higher order constructs about knowledge and reality, which are housed within paradigms (discussed in Section 1), with the more pragmatic techniques of research known as methods. A way of seeing this is as a *bridge* between abstract concepts and the reality of conducting research. If decisions are made which correctly align paradigms, methodology and methods, the study will be '*methodologically congruent*'. Methodological congruence is an important factor in study validity and refers to the coherence, compatibility or *fit* between the parts of a study. For example, the research question must *fit* the problem it seeks to address; the epistemology and ontology must *fit* with the research question; and the methodology of choice, together with the methods, must *fit* the paradigm leading the study.

HOW DOES THE PARADIGM INFLUENCE THE METHODOLOGY?

The examples listed below show three ways that that methodology can be informed by a specific paradigm. The illustration also shows the subsequent selection of a research design and some appropriate methods.

Paradigms

Interpretivism	Positivism	Pragmatism
	↓	

Methodology

Qualitative	Quantitative	Mixed

Research designs

Ethnography	Randomised control trial	Mixed methods
	↓	

Methods

Small purposive sample	Large sample	What works*
Observations	Randomisation	Questionnaires
Thematic analysis	Outcome measures	Interviews
Reflexivity	Inferential statistics	Triangulation

* In this context 'what works' means the researcher can choose any methods that enable them to answer their question. For example, they may ask a small number of people to take part in an interview and combine this information with that obtained from a questionnaire sent out to a much larger sample.

METHODOLOGICAL CONGRUENCE AND THE FINAL OUTCOME

A study is methodologically congruent if the final result represents the intended outcome of the chosen methodology. Where methods are used which are not appropriate for the methodology the final outcome is unlikely to reflect the methodology of choice or address the primary aim of the study. For example, if we go back to the cake analogy, if you want to make a chocolate gateau but change all the ingredients you will make something but probably not a cake. If you at least keep the core ingredients but add something new or change the way the ingredients are put together, you may still achieve a nice tasting cake, but it is unlikely that you will have the cake you intended. If you follow the recipe exactly as stated, you should end up with the cake you set out to make although, much like research, experience and familiarity of technique will always affect the final outcome.

CREATIVITY AND ORIGINALITY

While following instructions is important, there is room for creativity and originality. However, the methodology, its underpinning philosophy and the methods most commonly associated with it must be comprehensively understood first. Some methods are not flexible at all, while others may well be revised in certain contexts. For example, a quantitative methodology might use a small sample where a large sample is simply unfeasible. Similarly, a qualitative study may use a large sample if a team of researchers can still analyse the data in the depth and complexity expected for this methodology. Originality in research is exciting but novice researchers would be wise to stick closely to the methods prescribed.

KEEP IT SIMPLE

It is often said that the best research is the simplest. Such simplicity is obvious when studies demonstrate clarity, congruence and fit.

It's time for you to do some reflective work. I'd like you to think about epistemology and ontology and what methodology you might be more closely aligned to.

- Do you think knowledge is fixed?

- How can knowledge be multiple?

- What is important about objective truth?

- Why is subjective truth relevant?

- Do you think at this stage that you might like qualitative, quantitative or mixed methodology more? Why is this?

Section 3

How do I choose between quantitative, qualitative and mixed methodologies?

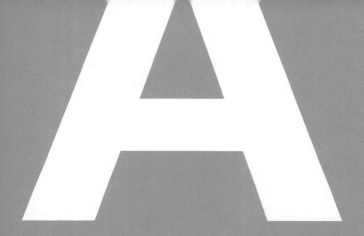

summary

Lots of things influence methodological
decision making including the research
question, feasibility, ethics and the
researcher themselves.

It all starts with a question

Research questions are very important in research and if you look closely you will often see descriptive words that reveal the researcher's methodological preference. However, even before the research question is set decisions are made about what can and cannot be investigated from both an ethical and a practical perspective. In addition to this the researcher will often have a methodological preference (think about the last checkpoint activity) which is influencing how they see the problem and how they want to design the study. Understanding the main commitments underpinning qualitative, quantitative and mixed methodologies will help you select an appropriate methodology for the study.

Quantitative studies reduce the world to things that can be measured or described using numbers.

Core commitments include:

- Being concerned with How often? How many? How much? and Does this work?
- Finding facts that can be externally checked and verified
- Having large sample sizes
- Describing patterns or looking for relationships in the data
- Being reproducible
- Being explicitly relevant for people beyond the study sample

CHOOSING QUANTITATIVE METHODOLOGY

In contrast, qualitative researchers often want to investigate experience, context and culture and understand how people make sense of the problem being investigated.

Core commitments include:

- Exploring topics in depth and in detail
- Collecting rich, complex data
- Using smaller samples because this depth and richness cannot be achieved without them
- Not assuming what is important before data collection
- Keeping an open mind about what might be found

CHOOSING QUALITATIVE METHODOLOGY

Researchers choose mixed methodology when the answer to their research question would be incomplete if they only collected data which fitted quantitative or qualitative methodologies. Mixed methodologists have therefore more freedom and more techniques available to them. However, do not think this is an easy option – far from it. The underpinning philosophies and theories guiding qualitative and quantitative research are so different it is really very hard to ignore them. Think about what happens to oil and water when placed in a bottle: they naturally want to separate. The same applies to qualitative and quantitative research because it is hard to reconcile these differences within one study. However, researchers can and do ignore much of the philosophical divergence between qualitative and quantitative methodologies and simply focus on what is best for the research question. This is hard but achievable. Think about the oil and water metaphor – if the bottle is shaken really vigorously it will make an emulsion. This is the same for mixing methodologies.

Core commitments include:

- Using what works for the research question
- Being pragmatic about competing philosophical positions
- Drawing together findings from both approaches to form a final conclusion

CHOOSING MIXED METHODOLOGY

Is one methodology better than another? The simple answer is no. One methodology is not more or less valid than another. Validity should be judged by the appropriateness of the methodology to the research question. For example, if you want to know how to reduce pain you need a quantitative study measuring pain which can clearly identify whether pain is reduced or not. If you want to understand the lived experience of chronic pain, you need to talk to people who have this experience and find out how their lives have been affected. This latter study would be qualitative. If you want to find out how a subjective understanding of pain can predict who will experience more acute pain this study might be mixed.

IS ONE METHODOLOGY BETTER THAN ANOTHER?

A well-formulated research question should reveal its methodological orientation. These implicit references reveal how the researchers are framing their investigation and what they think is important. Quantitative questions typically want to investigate cause and effect, correlation or describe the presence of well-defined variables in specific populations like satisfaction (think 'How satisfied are you on a scale of 1–5?). Look out for the following descriptors which may indicate that the study is quantitative.

- Cause
- Effect
- Correlation
- Description

- Reduce
- Increase
- Significant
- Improve

- Explain
- Predict

METHODOLOGY AND THE RESEARCH QUESTION

In contrast, qualitative studies pose broader questions often relating to experience, culture and or context. These questions are less tightly bound around outcomes, investigating the processes involved in these experiences instead. Look out for the following descriptors which may indicate that the study is qualitative.

- Impact
- Experience
- Meaning
- Understand
- What
- How
- Making sense
- Explore

Research questions in a study using mixed methodology will combine elements from both these lists. Depending on the way the study is designed there may first be a qualitative question and then a quantitative question to investigate (or of course vice versa). A good question in this context will reflect the intention of combining research approaches and how this will advance understanding, e.g. How does understanding an experience in more depth help predict who will benefit most from an intervention?

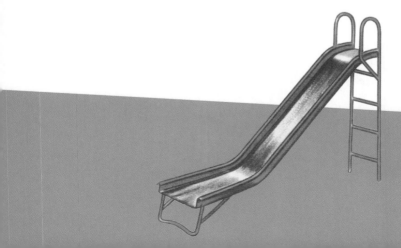

While methodology must always be chosen because it responds most appropriately to the needs of the research question, there is often an underlying influence from the researcher: the type of research they prefer, their values and their experience. These values will influence the study right from the formulation of the research problem and how the aims for the study are decided. Instead of trying to hide these, researchers are encouraged to recognise, critically reflect on, and show how these values are shaping the study. This is why it is hard, but of course not impossible, for a qualitative researcher to lead a purely quantitative study and, perhaps even harder, for a quantitative researcher to lead an in-depth qualitative study. It is also often difficult for either qualitative or quantitative researchers to design balanced mixed methodology studies because their preference will often influence how they prioritise the different types of data. The technical term for exploring how values shape the research design is *axiology*.

THE INFLUENCE OF THE RESEARCHER

Are these questions qualitative, quantitative or mixed methodology studies?

1. In what way does type 1 diabetes affect the lives of primary-school-age children?

2. Does acupuncture relieve symptoms of nausea for patients undergoing chemotherapy?

3. How satisfied are students with their research methods module and what influences their experience?

4. Does job satisfaction and work-related stress affect psychological wellbeing?

5. How does culture affect the experience of post-traumatic stress disorder following conflicts in developing countries?

6. How do people with autism spectrum disorder adjust to the working environment?

7. How, when and why do children run away from home?

Answers

1 Qualitative. 2 Quantitative. 3 Mixed. 4 Quantitative. 5 Qualitative.
6 Qualitative. 7 Mixed.

47

#LittleQuickFix

1

Section

When should I use quantitative methodology?

A

summary

When you want to find
a definitive answer to
How many? How much?
How often?

Definitive answers

Quantitative methodology is used when the researcher wants a definitive answer to a research question. This sharp focus on How many? How much? How often? means that quantitative research questions are typically very narrow. Researchers use quantitative research because they want to focus on specific issues which they can measure. Measurement can involve things like height and weight or opinions by using scales. Other data can be obtained, like gender, socioeconomic status and ethnicity. All this information is either already a number or can be converted to a number (e.g. Male = 0, Female = 1). Therefore, all quantitative methodology relies on numbers as its basic unit of analysis. This type of analysis means that a large sample can be included which makes the analysis more powerful and the final answer more definitive.

THE BENEFIT OF USING NUMBERS IN RESEARCH

The ability to *measure* the phenomena of interest is a fundamental part of quantitative methodology. Measurement produces numbers which are fixed, objective and verifiable. For example, if we measure height using centimetres on a ruler, if the ruler is accurate the height recorded from that ruler should be accurate too. If we are concerned about the accuracy of the person measuring height, we can ask a second person to check and if both people get the same answer, we can be more confident that the measurement is accurate. Researchers use tools to measure things like depression, stress, anxiety, quality of life, satisfaction. All these tools produce numerical results indicative, for example, of the amount of stress present in the person completing it.

QUANTITATIVE TOOLS

Quantitative researchers use 'tools' to collect their data. This is essentially a list of information they want to gather which might include something the participants already know about themselves such as age, gender, ethnicity, employment status, smoking status. Alternatively, the researchers may want to use the research to measure something more complex like quality of life or satisfaction, or they may want to identify the presence and intensity of something like depression. These latter examples can often be examined by using specific types of questionnaires called *outcome measures*. Good outcome measures have already been tested to ensure they do actually measure the thing they are supposed to (this is called *validity*) and that they can be used consistently with different people (this is called *reliability*).

Here are some examples.

Family relationship scales

- Family Assessment Device
- Family Adaptability and Cohesion Evaluation Scale
- Self-Report Family Inventory
- Family Alliance Assessment Scales

Satisfaction scales

- Satisfaction with Life Scale
- Student's Life Satisfaction Scale
- Marital Satisfaction Scale
- Job Satisfaction Scale

Quality of life measures

- McGill Quality of Life Questionnaire (MQOL)
- Health-related Quality of Life (HRQOL)
- WHOQOL
- EQ-5D
- Physical Quality of Life Index

KEY FEATURES OF QUANTITATIVE METHODOLOGY

If you want to find out the average age of unemployment you are more likely to determine an accurate estimate based on 3,000 participants than you are on 30. Quantitative research relies on large samples where the answer is directly relevant to those outside of the study sample. Making predictions beyond the study sample is called *generalisability* and is a core motivating factor for those wanting to conduct quantitative research. Another benefit of using numbers is that they can be analysed using *statistics*. Statistical analysis ranges from simple averages to more advanced calculations like T-tests. Each statistical test converts individual data to summary statistics like means, percentages or P values. These summary statistics are more useful as they represent the entire data set which can then be used to infer relationships in the wider population.

WHAT TYPES OF QUANTITATIVE METHODOLOGIES ARE THERE?

There are several types of quantitative methodology, ranging from experimental to observational designs.

Experimental studies investigate *cause and effect* by introducing something new to the participants, like a new treatment or intervention (as in randomised control trials).

Observational studies include descriptive and correlational designs which investigate the phenomena of interest by examining naturally occurring events.

- Descriptive studies are important to identify prevalence and incidence rates like opinions, attitudes, behaviours, morbidity and mortality (as in descriptive surveys).
- Correlational studies examine the patterns between things (*variables*) like job satisfaction and staff retention, education and social mobility, smoking and cancer (e.g. cohort/case control).

Sometimes researchers would like to have used an experimental methodology because they are really interested in cause and effect (like smoking and cancer), but ethics or feasibility prevents them from doing so. Therefore, the researchers have to use an observational design instead.

Answer the following questions to determine the appropriate methodology for your research.

Are you interested in How much? How often? How many? **YES / NO**

Could the answer to the research question be presented numerically? **YES / NO**

Can the phenomenon of interest be measured? **YES / NO**

Do you want to find a definitive answer? **YES / NO**

Do you anticipate using a large sample? **YES / NO**

Do you want to be able to generalise your result? **YES / NO**

If you have answered 'yes' to most of the questions above, then you should be using a quantitative methodology to answer your research question.

Section

5

When should I use qualitative methodology?

A

summary

When you need a rich,
in-depth study that can help
answer How do ... ? Why
do ... ? What is ... ? questions
which offer a more detailed
understanding of the problem.

Exploration and understanding

Qualitative methodology should be chosen when the research question is broad and exploratory in nature. Qualitative researchers aim to interpret, understand and illuminate the *subjective* experiences of participants. This emphasis on subjectivity is the cornerstone of qualitative methodology and suggests that experience should not be reduced to specific things that can be measured. Typically, qualitative questions are framed using How? Why? What? These questions are important because they reveal the reality of living in certain circumstances and the process of adaptation or adjustment, they help us understand context and why something works or does not. For many, qualitative studies provide an insight into reality that cannot be achieved by using numbers. In this way qualitative approaches are considered to be holistic and person-centred.

Qualitative methodology provides an opportunity to examine complex problems in depth and in detail. Qualitative researchers do not think numbers are an effective way to capture the human experience. Qualitative methodology does not aim to aggregate experience; instead it makes room for multiple realities to co-exist. Answers to qualitative questions are complicated, messy and often very personal. The researcher's role in qualitative methodology is to interpret the individual experience.

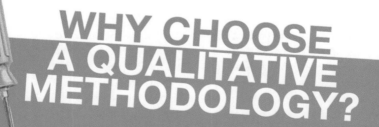

WHY CHOOSE A QUALITATIVE METHODOLOGY?

The idea of multiple realities is often difficult to make sense of. However, think about your age. Your chronological age is a definitive number, fixed and verifiable, but do you always feel, or act, your age? Do you feel older or younger depending on the context or environment you are in? In this example the experience of age is fluid: I feel old today and young tomorrow. The interesting thing for a qualitative researcher is to find out about the context and experiences that lead to people feeling differently in different circumstances. That is what makes qualitative research so rich.

MULTIPLE REALITIES

Qualitative researchers commit to in-depth, rich and complex investigations. Sample sizes must be kept small, so researchers have the opportunity to explore the data in sufficient depth. Good qualitative research takes time, and a great deal of effort. Data from interviews, observations, focus groups must be systematically broken down, compared, contrasted, described, interpreted, synthesised and rebuilt into something useful. Findings are often reported as *themes*. While it may not be possible to verify these findings definitively there are steps that qualitative researchers can take to increase their confidence that they are accurate. This is called *credibility*.

KEY FEATURES
OF QUALITATIVE
METHODOLOGY

These steps decrease the likelihood of naive and superficial interpretations of the data collected:

- Asking participants to confirm if the findings are representative of what they said (respondent validation)

- Asking another researcher to check the analysis and interpretations of the data (peer review)

- Spending more time with participants, such as using follow-up interviews (prolonged engagement in the field)

- Keeping a reflexive diary (audit trail)

STEPS TO INCREASE CREDIBILITY

There are many different types of qualitative methodology (in this text they are termed research designs), each trying to achieve something slightly different. Many will look and feel very similar, especially to a novice researcher. So, spending time exploring the nuances of each is time well spent when designing a qualitative study. The five most common types of qualitative methodology are listed below with a very brief descriptor of what this approach aims to achieve.

1   Ethnography (the study of culture)

2   Phenomenology (the lived experience)

3   Narrative inquiry (stories and storytelling)

4   Grounded theory (generating theory)

5   Case study (a bounded system)

WHAT TYPES OF QUALITATIVE METHODOLOGIES ARE THERE?

These five types are very specific research designs, and all have additional commitments beyond those described (on page 41) for qualitative research which set them out as unique. These commitments originate from their philosophical, epistemological and ontological roots. These approaches are immersive, extremely detailed, in-depth and can be very time-consuming. However, not all qualitative researchers want to sign up to these commitments and desire more freedom to design their study in a way that suits their aims. These studies are referred to as generic, descriptive, pragmatic, and make no claim to align their study to any specific philosophical perspective. You will see lots of these examples in the literature often simply described as 'qualitative research'.

- An ethnographic study of life inside a high-security prison

- The lived experience of gender reassignment

- Storied accounts of living with chronic illness and identity reconstruction

- A theory of family adjustment following divorce

- Striving for excellence: A case study of one school's experience of educational reform

- A qualitative analysis of student experiences during their first year at university

SOME EXAMPLE TITLES OF QUALITATIVE RESEARCH

SHOULD YOU USE QUALITATIVE METHODOLOGY?

Answer the following questions to determine the appropriate methodology for your research.

Are you interested in How much? How often? How many?   **YES / NO**

Could the answer to the research question be presented numerically?   **YES / NO**

Can the phenomenon of interest be measured?   **YES / NO**

Do you want to find a definitive answer?   **YES / NO**

Do you anticipate using a large sample?   **YES / NO**

Do you want to be able to generalise your result?   **YES / NO**

If you have answered 'no' to most of the questions you will most likely want to use a qualitative methodology.

When can I mix qualitative and quantitative methodologies?

A

summary

When the research question
cannot be fully answered by
either qualitative or quantitative
methodology alone.

Why not just do a bit of both?

It may seem a really good idea to mix methodologies in circumstances where the researcher can see the benefit of both qualitative and quantitative approaches to address their research problem. However, the process of exploring the research problem, and of course reviewing the associated published literature, will very often lead the researcher towards a research question which can and should be answered entirely by either qualitative or quantitative methodology. However, when the problem, and its question, do not frame themselves in this way, and **if using only qualitative or quantitative approaches the question would be left partially answered, then mixing methodologies may be a very good solution.**

WHAT ISN'T MIXED METHODOLOGY?

It might seem odd to start with what isn't mixed methodology rather than what is but clarity on this point will help you understand what proper mixed methodology is. For example, it is possible that in a quantitative study you need a little bit of qualitative data like adding an open question to an otherwise closed answer questionnaire. However, open questions in a questionnaire are often analysed with statistics, as answers are grouped and counted to analyse for trends. So, a study like this does not attempt any qualitative analysis of the open questions.

Alternatively, in a qualitative study you may want to collect some quantitative data like age, gender, occupation. However, a study like this would not analyse the quantitative data to identify trends or patterns in a way that a quantitative study would normally do. Instead, this type of data is collected to deepen the analysis and increase accuracy in the interpretation of the qualitative data. Therefore, when researchers simply add a small amount of different types of data to their study they usually do not think too hard about the philosophical reasoning about why this may or may not be appropriate, and often do not analyse this additional data in much depth. Therefore, such studies are not mixed methodology.

SO, WHAT IS MIXED METHODOLOGY?

In contrast to the examples given above, researchers who use mixed methodologies really use the data drawn from qualitative and quantitative approaches in detail by applying appropriate analysis techniques to draw conclusions. There are lots of ways in which studies can combine qualitative and quantitative approaches. Here are just a few examples:

1   A large quantitative survey followed by some individual interviews

2   A focus groups study used to inform the development of a questionnaire

3   A study which uses interviews, questionnaires and observations concurrently

In the first two examples one approach is completed first and the second approach takes place afterwards (known as sequential mixed methodology). For example, in a study which relies on survey data the answers can be superficial. So, if the researchers want to know a little more about the problem, they might use the results of the survey to identify people to conduct an in-depth interview with. Similarly, researchers may use a focus group to explore a topic first, analyse the data and use it to create a questionnaire that can be sent out to more people so they too can understand the problem in more depth. In the third example researchers may conduct a number of different data collection techniques broadly at the same time, analyse them individually and then draw all this information together in a final step called *triangulation*. Triangulation is a process of directly comparing data from qualitative and quantitative approaches to reach a more accurate conclusion.

WHAT TYPES OF MIXED METHODOLOGIES ARE THERE?

There are numerous different types of mixed methodology and it is easy to feel overwhelmed when reading about this approach. Therefore, only three types are introduced to you here. However, if you want to design a mixed methodology study you would benefit from spending time examining other types in more detail. These three types vary in the importance they place on the different types of data in the study. This is because when mixing methodologies it is very common that – despite valuing both data types – one is given priority over the other. In this case there is a leading methodology on which the study is based, and the other data collected is treated more as a supplement. We might see this in a large quantitative study which uses a follow-up qualitative interview to collect further in-depth information. Similarly, a researcher might conduct a small and in-depth qualitative study but add a structured questionnaire to further understand the participants' experience. In contrast, there are studies that treat the data from both approaches equally.

These three approaches are illustrated below

Type	Leading methodology
Big QUANT – little QUAL	Findings are predominantly drawn from the quantitative data but the qualitative aspect of the study helps to add further insight
Big QUAL – little QUANT	Findings are predominantly drawn from the qualitative data, but the quantitative data helps understand this more clearly
QUANT – QUAL	Findings from both quantitative and qualitative data are treated equally to determine the answer to the research question

Are the following examples
of mixed methodologies?
Answer yes or no.

1. A survey about adjustment to motherhood which uses scales (strongly agree–strongly disagree) with a few questions that allow the participant to add their own comments? Yes / No

2. A study on the effect of immigration on a rural community which collects data on employment, deprivation, satisfaction, wellbeing and also interviews local residents, employers and school teachers Yes / No

3. An interview study of living with diabetes which ask participants for their age, gender, ethnicity, weight and height Yes / No

4. A study that uses focus groups to explore students' perceived stress during their undergraduate programme and then designs and distributes a questionnaire based on the answers in the focus group Yes / No

5. A study that wants to find out if a new diet increases weight loss and then interviews participants about their experience of compliance with the new dietary regime Yes / No

No – this is a quantitative study which may simply review responses from open questions and categorise them rather than applying any qualitative analysis

 Yes – this study will collect quantitative and qualitative data simultaneously and analyse both to answer the research question

 No – this is a qualitative study which needs some background information to understand the people participating in the study

 Yes – this study is a two-phase mixed methodology study using the qualitative data to inform the quantitative data collection phase

 Yes – this is a two-phase mixed methodology study. The first phase will use statistical analysis to look at the treatment effect of the new diet and then follow up with a qualitative study to look at participants' experiences to understand compliance in more depth

CONGRATULATIONS! YOUR PREFERRED METHODOLOGY IS

. . . .

SECTION

7

DIY: HOW DO I CHOOSE
THE RIGHT METHODOLOGY
FOR MY RESEARCH?

KNOW YOUR PROBLEM,

KNOW YOUR QUESTION,

KNOW YOURSELF

Choose your methodology

This DIY section will help you choose a methodology for a study you want to design. There are three stages to selecting a methodology. Once you have properly worked through them you should know which to select.

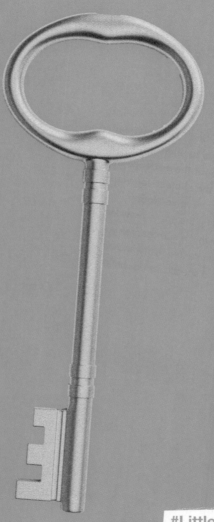

KNOW YOUR PROBLEM

The first thing you need to do is to understand more about the evidence base of the subject you want to investigate.

Using the space provided, jot down the subject area you are interested in (try and be as specific as possible).

1

My topic area is:

Other relevant details about my topic include.

Now follow the steps below.

1 Open a journal database (if you are not sure what database to use, talk to a librarian to guide you through the most appropriate one for you subject)

2 Conduct a search for your topic of interest

3 Make a note of interesting titles that are within your topic area

4 Now use the table on the next page to record the title, research question and methodology for ten of the most relevant papers to your topic area

	Title	Question	Methodology
1			
2			
3			
4			
5			
6			
7			
8			
9			
10			

5 Now reflect and write down …

What could a new qualitative study contribute to this evidence base?

What could a new quantitative study contribute to this evidence base?

What could a new mixed methodology study contribute to this evidence base?

6 Given your answers above, which methodology would make the strongest contribution to the evidence base?

KNOW YOUR QUESTION

The second thing you need to do is understand what type of research question you want to develop.
(You may want to read the *Little Quick Fix: Research Question* to help you further with this.)

2

Review the descriptors listed and circle those that most reflect the sort of question you want to set.

Impact	Meaning	Correlation	Explanation
Cause	Experience	Making sense	Prediction
Effect	Description	Exploration	How do?
Understanding	Reduction	Significance	What are?
Increase		Improvement	

If you have circled mostly orange this will be a qualitative question; if you have circled mostly blue this will be a quantitative question. If you have circled several oranges and blues you either haven't made up your mind yet, or you may want to set a mixed methodology question.

KNOW YOURSELF

Time for some reflective work again. Having read about qualitative, quantitative and mixed methodologies, do you gravitate towards liking one of these more than the other?

3

In the space below, make a note of what you have read, what you have done and what you have been taught that has influenced this preference.

Here are some example reflections.

'Lectures on statistics were too hard'

'Love the idea of exploring a topic in-depth'

'Don't believe you can learn anything with small samples'

'Questionnaires are too superficial to learn anything useful'

To be confident that you know how to choose your methodology work through this checklist.

☐ Can you define methodology and identify the epistemology and ontology for qualitative, quantitative and mixed methodologies? If not, go back to pages 13 and 15.

☐ Do you understand why methodology is important in research and what type of research you might have a preference for? If not, go back to pages 27 and 33.

☐ Are you able to identify methodological commitments from descriptors in research questions? If not, go back to pages 44–45.

HOW TO KNOW
YOU
ARE
DONE

CHECKPOINT

☐ Do you understand when a quantitative methodology would be most appropriate for a study? If not, go back to page 61.

☐ Do you understand when a qualitative methodology would be most appropriate for a study? If not, go back to page 75.

☐ Do you understand what is and what isn't considered mixed methodology? If not, go back to pages 82–84.

Glossary

Axiology The values of the researcher influencing the study

Congruence Compatibility or 'fit' between all aspects of the research design

Credibility Confidence that the findings of a qualitative study are accurate

Epistemology The nature of knowledge and how knowledge can be gained

Generalisability Being able to assume that the conclusions of a study are applicable to those outside of the study sample

Interpretivism A paradigm for qualitative research that commits to the interpretation of individual experiences

Methodology The fundamental assumptions that guide decision making in a study

Ontology The nature of truth and reality

Outcome measure Types of questionnaire validated to test a particular outcome

Paradigm A world view consisting of philosophy about what is influencing the conduct of the research

Positivism A paradigm for quantitative research which assumes a fixed and objective reality

Pragmatism A paradigm for mixed methodologies allowing pluralistic methods of data collection

Qualitative Research approach that investigates experience, context and culture

Quantitative Research approach that reduces the world to things that can be measured or described numerically

Statistics A calculation made using numerical data

Triangulation Use of multiple methods to converge on a more accurate interpretation of the data